REINDEER

Amazing Facts About Nature's Arctic Travelers for Kids

Dylanna Press

Copyright © 2026 by Dylanna Press
Author: Tyler Grady

All rights reserved. No part of this publication may be reproduced, stored in a retrieval system, or transmitted by any means, including electronic, mechanical, photocopying, or otherwise, without prior written permission of the publisher.

Although the publisher has taken all reasonable care in the preparation of this book, we make no warranty about the accuracy or completeness of its content and, to the maximum extent permitted, disclaim all liability arising from its use.

Trademarks: Dylanna Press is a registered trademark of Dylanna Publishing, Inc. and may not be used without written permission.

ISBN: 978-1-64790-446-3
Publisher: Dylanna Publishing, Inc.
First Edition: 2026
Printed in the United States of America
10 9 8 7 6 5 4 3 2 1

For information about special discounts for bulk purchases, please contact:

Dylanna Publishing, Inc.
www.dylannapublishing.com

Contents

Meet the Reindeer   7
Built for the Arctic   8
Where Do Reindeer Live?   11
Super Survivors: Reindeer Adaptations   12
What Do Reindeer Eat?   15
Life in the Herd   16
On the Move   19
A Day in the Life   20
Mating and Birth   23
Growing Up Reindeer   24
Reindeer and Their Ecosystem   27
Natural Predators   28
Challenges and Threats   31
Life Span and Population   32
Conclusion   35
Test Your Reindeer Knowledge!   36
STEM Challenge: Think Like a Scientist!   37
Glossary   39
Resources and References   40
Index   41

Fun Fact: A reindeer's eyes change from golden brown in summer to blue in winter!

Meet the Reindeer

CLOMP! Picture thousands of animals moving together across a frozen landscape, their hooves clicking in rhythm like a drumbeat you can hear from miles away. The ground trembles beneath their weight as an endless river of antlers flows toward the horizon. Welcome to the world of reindeer!

These powerful animals survive in some of the coldest, harshest places on Earth. You'll find them roaming the **Arctic** regions of Alaska, Canada, Greenland, Scandinavia, Russia, and Siberia, where winter temperatures plummet to –40°F (–40°C) and howling winds can freeze exposed skin in minutes.

Their scientific name is *Rangifer tarandus*. In North America, wild populations are called **caribou**, while the semi-domesticated herds in Europe and Asia are known as reindeer.

Reindeer vs Moose

Reindeer	Moose
Reindeer live in the tundra and northern forests of Europe, Asia, and North America.	Moose live in the forests and wetlands of North America, Europe, and northern Asia.
Both male and female reindeer grow antlers each year.	Only male moose grow antlers, which are broad and flat.
Reindeer travel in large herds and migrate long distances each year.	Moose are solitary animals, usually seen alone or with a calf.
Reindeer have wide, snowshoe-like hooves for walking on snow.	Moose have long legs and pointed hooves for wading through water.

What makes them so incredible? Some herds travel up to 3,000 miles (4,800 km) each year—that's like walking from New York to California! They have special hooves that change with the seasons, thick coats with hollow hairs that trap warmth like a thermos, and eyes that change color from golden brown to deep blue. Each of these features helps them survive where most animals would freeze.

For thousands of years, Arctic peoples have depended on reindeer for food, clothing, shelter, and transportation. Today, these animals still thunder across the frozen north—some in wild herds, others alongside traditional herders—but their numbers have fallen sharply. Many herds that once had hundreds of thousands of animals now have only a few thousand left.

Climate change, roads, mines, and other human activities are making life harder for reindeer every year. Scientists and **Indigenous** communities are working together to protect them because losing the reindeer would mean losing a vital part of the Arctic itself, and a living link to our planet's wild past.

Built for the Arctic

Reindeer are big animals! Males (called bulls) weigh between 350 to 400 pounds (160 to 180 kg)—about the same as a small horse. Females (called cows) are smaller at 180 to 260 pounds (80 to 120 kg). They stand about 4 to 5 feet (1.2 to 1.5 meters) tall at the shoulder, roughly the height of a tall refrigerator.

Their antlers are spectacular. Male reindeer grow huge antlers that can spread 4 feet (1.2 meters) wide and weigh up to 33 pounds (15 kg). That's like carrying a bowling ball on your head! Female reindeer also grow antlers—they're the only deer species where females have them—though theirs are usually smaller.

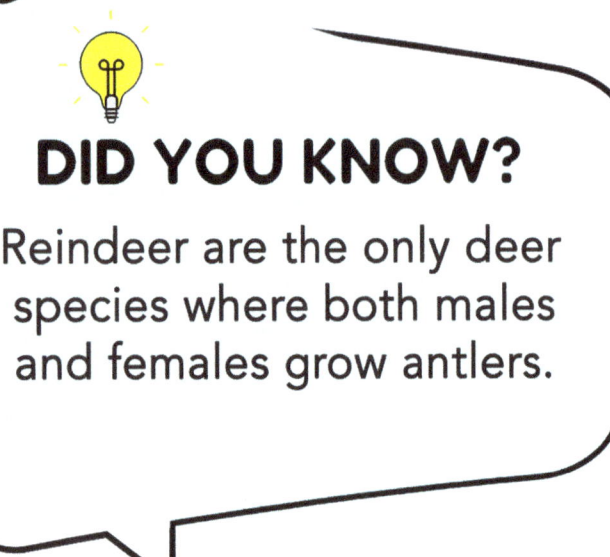

DID YOU KNOW? Reindeer are the only deer species where both males and females grow antlers.

A reindeer's coat is like wearing the world's best winter jacket. They have two layers: fluffy wool close to their skin, and longer hollow hairs on top. Those hollow hairs trap air bubbles, creating insulation so good that reindeer can lie on snow without melting it. In summer, they shed this thick coat and grow shorter, darker brown fur.

But their hooves might be their coolest feature! These amazing feet change with the seasons. In winter, the soft pads inside the hooves shrink, exposing sharp edges perfect for gripping ice. In summer, the pads expand and get spongy for walking on wet, muddy ground. The wide, crescent shape spreads their weight like snowshoes, and they even help reindeer swim across rivers!

Reindeer have short, powerful legs and broad noses covered with fur to warm the freezing air they breathe. Their sense of smell is incredible—they can sniff out food buried 2 feet (60 cm) under the snow! Their hearing is excellent too, helping them detect predators.

Here's something wild: reindeer eyes change color! In summer, their eyes look golden-brown, but in winter they turn blue. This helps them see better during the months of Arctic darkness. They can also see ultraviolet light, which makes things stand out against white snow.

Where Do Reindeer Live?

Reindeer live at the top of the world, in the Arctic and sub-Arctic regions where it's cold most of the year. Their home territory circles around the North Pole, from Alaska and northern Canada to Greenland, Norway, Sweden, Finland, and across northern Russia and Siberia.

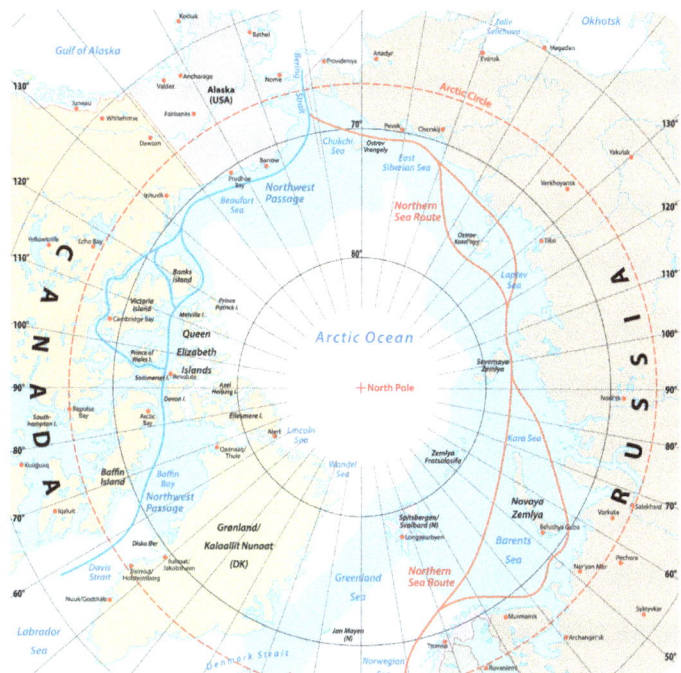

During the brief Arctic summer, many herds migrate to the wide-open **tundra**—a vast, treeless plain that stretches to the horizon like a frozen sea. Here they feast on grasses, flowers, and shrubs, building up fat for winter.

When winter comes, many herds migrate south to the **boreal forest** (also called taiga). These evergreen forests provide some shelter from howling winds and have lichens growing on the ground and trees. Other reindeer stay on the tundra all year, digging through snow to find food.

Some reindeer live in mountains, spending summers in high, cool areas and moving to lower valleys for winter. Others live near the coast, where they can eat seaweed and enjoy slightly warmer ocean air.

In North America, there are different types of caribou. Barren-ground caribou make the longest migrations across the tundra. Woodland caribou live in forests. Mountain caribou live in—you guessed it—mountains. And Peary caribou live on the northernmost Arctic islands where almost nothing else can survive.

In Europe and Asia, you'll find wild reindeer in the mountains of Norway, plus huge herds that are semi-wild and watched over by people like the Sámi, Nenets, and Evenki peoples. These herders have cared for reindeer for countless generations.

Reindeer need different types of places at different times. They need summer **calving grounds** where mothers can have babies safely. They need paths to travel between seasons. They need winter areas where they can find food under the snow. When roads, pipelines, or buildings block these paths, reindeer struggle to survive.

The Arctic is changing fast. It's getting warmer, which sounds nice but actually causes big problems. Plants are changing, ice is melting at weird times, and rain sometimes falls in winter and freezes into sheets that reindeer can't dig through to reach food.

Super Survivors: Reindeer Adaptations

Reindeer have incredible features that help them survive where most animals would freeze:

- **Super Insulation:** Their fur is so good at trapping warmth that snow landing on a reindeer's back won't melt. Seriously! The hollow hairs create tiny air pockets that hold heat close to their body. Even in -40°F weather, reindeer stay toasty warm.

- **Shape-Shifting Hooves:** Reindeer hooves change with the seasons. In winter, the edges harden to grip ice and scrape through frozen snow. When the tundra turns wet and muddy in summer, the pads soften to give better traction. Their broad shape even helps them swim—part snowshoe, part paddle!

- **Built-In Heater**: Inside their nose is a special system that warms freezing air before it reaches their lungs. It also captures moisture from the air they breathe out, so they don't lose water. It's like having a heat recovery system built into your face!

- **Magic Eyes:** Reindeer can see ultraviolet light (which humans can't see at all), helping food and predators stand out on snow.

- **Clicking Hooves:** When reindeer walk, their leg tendons make a clicking sound—click, click, click, click. In a herd of thousands, it sounds like waves crashing on a beach! This helps them stay together in blizzards and darkness, like having built-in walkie-talkies.

- **Smart Antler Timing:** Male reindeer drop their antlers after mating season in fall. But females keep theirs all winter. This helps pregnant mothers push males away from food craters in the snow, making sure they eat enough for their growing babies.

- **Fat Batteries:** During summer's non-stop eating, reindeer pack on layers of fat under their skin and around their organs. These fat stores are like backup batteries, giving them energy when winter food is scarce and not very nutritious.

- **Super-Baby Development:** Newborn reindeer calves can stand up within an hour and outrun a human by day two! This lightning-fast development is crucial because predators are everywhere and the herd keeps moving.

These **adaptations** didn't just appear overnight. They developed over thousands of years of living in the Arctic.

Fun Fact: Each reindeer grows a brand-new set of antlers every year.

REINDEER MATH

> A hungry reindeer may dig 30 feeding craters in a single day to find buried lichens. If each crater takes 5 minutes to dig, how many total minutes does the reindeer spend digging every day?

A: 150 MINUTES, OR 2½ HOURS

What Do Reindeer Eat?

Reindeer are **herbivores** whose menu changes from season to season.

Summer Feast: When the snow melts and plants burst to life, reindeer go into eating overdrive! They munch on fresh grasses, leafy plants, flowers, mushrooms, and the tender leaves of willow and birch shrubs. A reindeer might eat 12 pounds (5.5 kg) of plants a day, spending 12-18 hours just eating and eating and eating. They're building up fat reserves like a bear preparing for hibernation. Young reindeer can gain several pounds per day!

Fall Shift: As plants die back and turn brown, reindeer switch to eating twigs, dried grasses, and whatever green plants remain. They start relying more on the fat they stored up during summer.

Winter Challenge: This is when things get really tough. The main winter food is **lichen**—a fuzzy, crusty organism that grows slowly on rocks, soil, and trees. (People sometimes call it "reindeer moss" but it's not actually moss!) Lichen stays nutritious under the snow, but there's not much energy in it. When food is scarce, some have been seen nibbling on bird eggs or small bones for extra minerals!

Here's the amazing part: reindeer can smell lichen buried up to 2 feet (60 cm) under the snow! Once they find it, they dig feeding craters with their front hooves, scraping away snow in circles. One reindeer might dig dozens of craters in a single day. It's hard work and doesn't provide much energy, so winter reindeer move less and lower their body temperature to save energy.

Reindeer have four stomach chambers (like cows) that help them squeeze every bit of nutrition from their food. They're also "**ruminants**," which means they chew their cud—they swallow food, bring it back up later, and chew it again. Gross but effective!

Climate change is making finding enough food harder. When winter rain falls and freezes, it creates ice layers that reindeer can't break through. Whole herds have starved because ice blocked their food. Scientists are finding more and more dead reindeer who couldn't reach the lichens buried below the ice.

Life in the Herd

Reindeer are social animals that live in groups ranging from small family bands to massive **herds** of tens of thousands. Living together helps them survive the harsh Arctic.

For most of the year, males and females live separately. Mothers, babies, and young reindeer form maternal herds. Adult males hang out in bachelor groups or wander alone. These groups only mix during **migration** and breeding season.

Herd size changes dramatically with the seasons. During summer migration, reindeer form enormous gatherings. The Porcupine caribou herd in Alaska and Canada can have nearly 200,000 animals—imagine a living river of reindeer flowing across the tundra! They all gather in traditional calving areas where conditions are best for newborns.

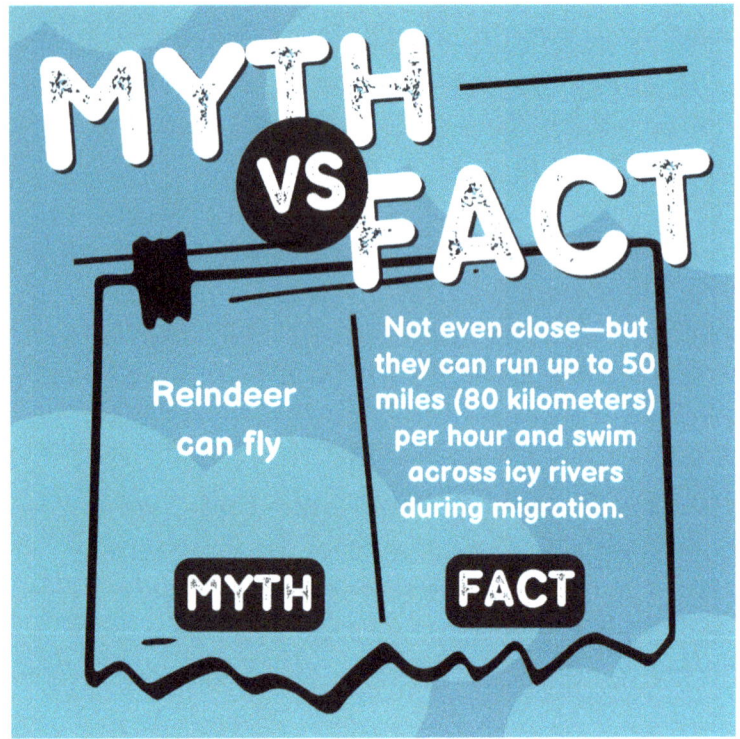

Why form such huge herds? Safety in numbers! With thousands of eyes watching, someone always spots predators. Plus, moving in a massive group helps them deal with the billions of biting mosquitoes and flies that swarm in Arctic summer. And being in the traditional calving area means babies are born where food is plentiful.

Reindeer "talk" in several ways. Mothers and calves recognize each other's unique calls—super important in a herd of thousands! Males grunt during breeding season. Alarmed reindeer snort to warn others. Their trademark clicking helps the herd stay together in fog and darkness.

There's a pecking order in reindeer herds based on body size, age, and antlers. But here's something cool: in winter, pregnant females with antlers boss around males without antlers! This helps mothers-to-be get access to food craters.

When predators show up, reindeer have different strategies. Sometimes they bunch together with calves in the middle. Sometimes they scatter in all directions to confuse the hunters. During migration, they take turns at the front, breaking trail through deep snow until they're tired, then someone else moves up.

In autumn, the herds merge into massive mixed groups for breeding season. After that, males and females mostly separate again for winter.

Fun Fact: Reindeer can travel 12–30 miles (20–50 km) per day during migration—and run up to 50 miles per hour (80 km/h) when chased!

On the Move

Reindeer are true travelers of the north, famous for their incredible migrations across the Arctic tundra and forests. These journeys are among the longest of any land mammal—some herds travel more than 3,000 miles (4,800 kilometers) each year in search of food and better conditions. The size of their range depends on the herd, habitat, and season, stretching from coastal plains to mountain valleys and dense boreal forests.

Within their vast territories, reindeer follow well-established migration routes that have been used for generations. These paths connect calving grounds, summer grazing areas rich in grasses and herbs, and winter feeding sites where lichen grows beneath the snow. As they travel, thousands of hooves pack down the snow and shape the Arctic landscape, leaving trails that can be seen from above.

Reindeer migration is closely tied to the changing seasons. In spring, herds move northward toward cooler regions filled with fresh vegetation, giving mothers a safe place to give birth. By autumn, they head south again to escape deep snow and find shelter from fierce winter storms. Timing is everything—moving too early or too late can mean the difference between plenty and hunger.

Communication keeps the herd coordinated on the move. Clicking sounds from tendons in their legs help individuals stay together in fog or blowing snow. Scent and body language signal alarm or reassurance, and experienced leaders often guide the group to traditional feeding grounds.

Even outside the major migrations, reindeer are almost always on the move—walking, grazing, or pawing through snow for food. Their endurance, memory, and cooperation make them some of nature's most remarkable wanderers, perfectly tuned to the rhythm of the Arctic seasons.

A Day in the Life

A reindeer's daily routine looks completely different in summer and winter.

Summer Days

Reindeer are most active in the cooler parts of the day—early morning, evening, and even through the bright Arctic night. (Remember, there's 24-hour daylight in summer!) The middle of the day is rest time, especially when mosquitoes and flies are at their worst.

A summer day begins with feeding. Reindeer move slowly across the tundra, nibbling on grasses, flowers, and shrubs. After several hours of eating, they lie down to chew their cud—bringing food back up from their first stomach to re-chew it for better digestion.

While adults rest, calves play nearby and mothers keep watch. There's always at least one reindeer standing guard. At the first sign of danger—SNORT!—everyone jumps up, ready to run.

On bad bug days, herds cluster tightly together, stand in water, or seek windy ridges and snow patches to escape biting insects. Some days are all about bugs, not food!

DID YOU KNOW? Reindeer don't sleep through the night—they take dozens of short naps every day!

Winter Days

Winter brings darkness and stillness. Reindeer conserve energy, becoming most active during the warmer midday hours. Their eyes adjust to the gloom, and their powerful hooves dig through snow to uncover food.

Much of the day is spent digging, eating, and moving slowly to the next spot. Between feeding sessions, they stand quietly or rest in sheltered areas, lowering their metabolism to save energy. Their coats are so insulating that snow piles up on their backs without melting—they look like walking snow sculptures!

Breeding Season

In autumn, bulls focus on mating. They fight rivals, make grunting calls, and herd females—often losing weight in the process. Cows keep their regular routine, sometimes avoiding the rowdy males.

All year long, reindeer stay alert. Even while feeding or resting, their ears swivel and their noses test the air. In the open tundra, staying aware is everything.

Fun Fact: When reindeer lie down, they often curl their legs underneath and rest their heads on their sides—almost like dogs taking a nap.

Mating and Birth

The reindeer breeding season happens from September through November. This timing means babies will be born the following spring when food is plentiful and weather conditions are less severe.

The Rut: This is what scientists call breeding season. Bulls become aggressive and focused, engaging in loud grunting contests, sparring matches where they lock antlers and push, thrashing bushes with their antlers to show strength, and herding females away from other males.

These competitions can be intense! Bulls might fight for hours. Bigger males with larger antlers usually win, but stamina and determination matter too. Most fights end with displays rather than serious injury—the weaker male backs down before anyone gets hurt.

Successful bulls might mate with many females. Bulls put so much energy into breeding that they barely eat and can lose 50 pounds (23 kg) or more!

After breeding, males and females separate again. Bulls' antlers fall off in late autumn or early winter, but pregnant females keep theirs all winter long.

Female reindeer are pregnant for about 7.5 months. As spring approaches they migrate toward calving grounds, driven by instinct.

Babies are born in late spring or early summer, typically May or June. This timing is critical—any earlier and there's not enough food; any later and calves won't grow big enough before next winter.

Mothers usually have one calf, though twins sometimes happen. Birth usually occurs away from the main herd in a quiet spot. Labor is quick, often less than an hour.

Newborns weigh about 13–20 pounds (6–9 kilograms) and are covered in soft reddish-brown fur that helps them stay warm in the cold spring air. Each calf is born alert and ready to follow its mother across the tundra's open spaces, where predators are always nearby.

NEWBORN CALF STATS
- Birth weight: 13–20 lb (6–9 kg)
- Standing: within 1 hour
- Walking: by 1 day
- Weaning: 5–6 months

A powerful bond forms right away between mother and calf. They recognize each other by scent and voice, an essential connection that allows them to reunite even in a herd of thousands. This strong attachment gives the calf its best chance of survival through the challenging Arctic seasons ahead.

Growing Up Reindeer

Life as a baby reindeer is all about growing fast and learning survival skills.

Lightning Start: Within one hour of birth, the wobbly-legged calf can stand. Within four hours, it can walk beside mom. By day two, it can outrun you! This super-fast development is critical—the herd won't wait, and predators are always watching. The calf stays glued to its mother, nursing on incredibly rich milk (22% fat!) that fuels explosive growth.

Turbo Growth: Calves gain 2-3 pounds every single day! That 15-pound newborn doubles its weight in just two weeks. By midsummer, calves weigh 60-80 pounds and have traded their reddish baby fur for a proper brown coat. They start nibbling plants while still nursing, copying everything mom does—what to eat, where to dig, how to spot danger.

Play Hard, Learn Fast: Young reindeer are incredibly playful! They race in circles, leap over rocks, and practice-fight with their tiny antler buds. This isn't just fun—it builds muscles, speed, and survival skills. Calves form friendships that can last years, learning who's brave, who's fast, and who makes a good ally in times of danger.

The Ultimate Test: A calf's first winter separates survivors from the rest. They must learn to dig through snow for food (though their small hooves aren't very good at it yet), find shelter from screaming winds, and stay alert for wolves—all while still growing. Only about half of all calves make it through their first year.

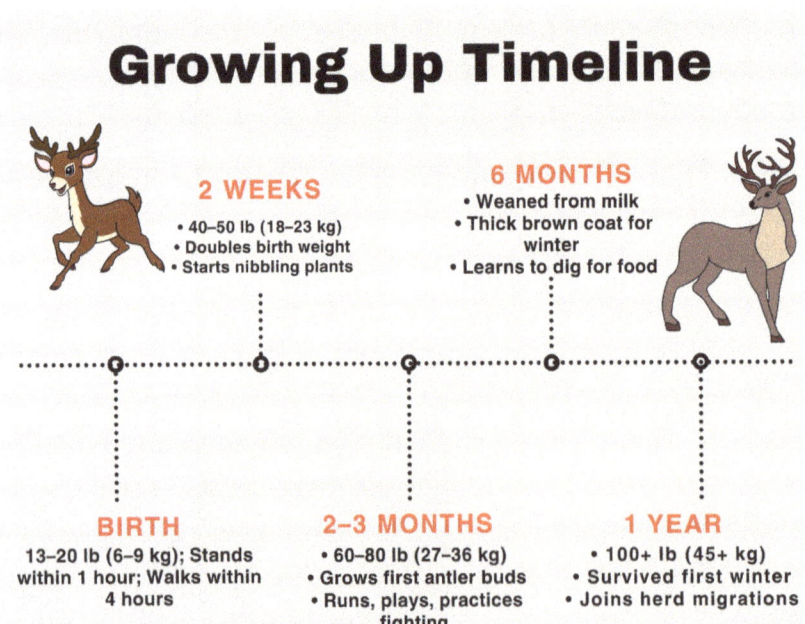

Growing Up Timeline

2 WEEKS
- 40–50 lb (18–23 kg)
- Doubles birth weight
- Starts nibbling plants

6 MONTHS
- Weaned from milk
- Thick brown coat for winter
- Learns to dig for food

BIRTH
13–20 lb (6–9 kg); Stands within 1 hour; Walks within 4 hours

2-3 MONTHS
- 60–80 lb (27–36 kg)
- Grows first antler buds
- Runs, plays, practices fighting

1 YEAR
- 100+ lb (45+ kg)
- Survived first winter
- Joins herd migrations

Growing Apart: Female calves often stay with mom for over a year, learning by watching. Males gradually drift away to join bachelor groups by age two or three. Both sexes grow their first tiny antlers in their first summer. Each year these fall off and regrow bigger, eventually becoming the impressive branches we recognize.

This harsh environment doesn't give second chances. Those calves that survive their first year prove they're tough animals ready for Arctic life.

Reindeer and Their Ecosystem

Reindeer aren't just survivors of the frozen north—they help shape the entire Arctic ecosystem. Every step, bite, and migration affects the land, plants, and animals around them.

Landscape Sculptors: As reindeer graze, they transform the land. Their nibbling keeps shrubs from taking over the tundra, keeping wide, open spaces for other animals. Their hooves churn the soil and help new plants grow.

Nutrient Movers: Reindeer are like walking fertilizer spreaders! They eat plants in one place and drop nutrients somewhere else. Over thousands of years, this has created rich feeding areas where new plants thrive. Calving grounds, used year after year, grow especially lush from this natural recycling.

Predator Lifeline: Many Arctic predators, like wolves and bears, depend on reindeer to survive.

Plant Keepers: Reindeer help keep the tundra a tundra! By eating shrubs and trampling young trees, they stop forests from spreading north. Their winter digging spreads seeds and mixes soil, shaping where different plants grow.

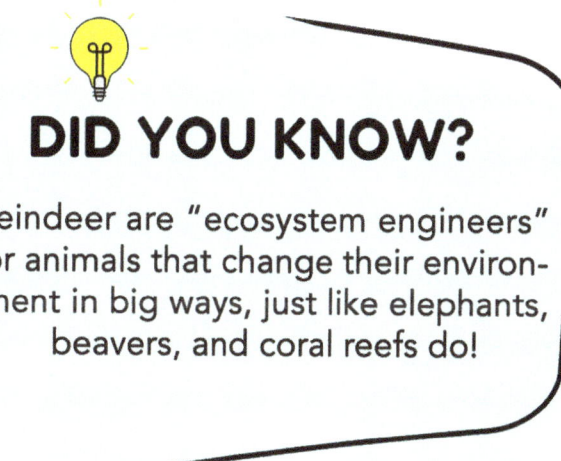

DID YOU KNOW?
Reindeer are "ecosystem engineers" or animals that change their environment in big ways, just like elephants, beavers, and coral reefs do!

Scavenger Support: Even in death, reindeer give life. A single carcass can feed dozens of Arctic animals like ravens, foxes, and even tiny shrews. In the freezing north, where every calorie counts, one reindeer can mean survival for many others.

Arctic Health Indicator: How reindeer herds are doing tells scientists about overall Arctic health. When reindeer struggle, the whole Arctic is struggling.

Protecting reindeer and their migratory routes means protecting the Arctic itself. These resilient animals link the health of the tundra, forests, and the many creatures, including humans, who depend on the stability of northern ecosystems.

Natural Predators

Life as prey animal means reindeer must stay constantly alert, ready to run at the first hint of danger. Even though they're big and strong, survival means staying alert, running fast, and sticking with the herd.

Wolves are among the most dangerous predators. Wolf packs often follow migrating caribou herds. In some places, wolves eat almost nothing but caribou. Their hunting strategies include chasing herds until a weak animal falls behind, setting up ambushes at river crossings, and coordinating attacks to separate individuals from the group.

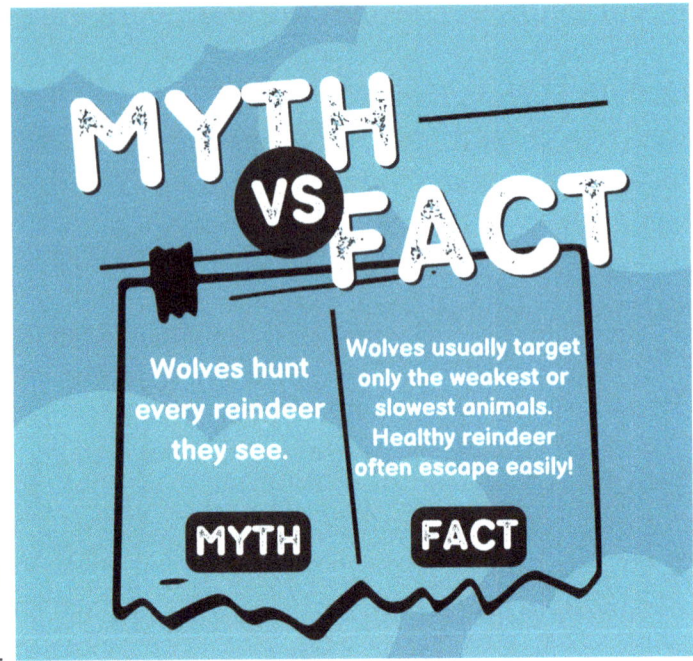

Bears are serious threats to baby reindeer. Brown bears and grizzlies actively hunt calving grounds in spring, looking for newborns who haven't developed their speed yet. A bear can kill multiple calves in one day. Polar bears in coastal Arctic areas sometimes hunt caribou too.

Other predators include **wolverines and lynx**, skilled solitary hunters that can ambush smaller reindeer or calves in forests and rocky terrain. In some regions, **golden eagles** have even been seen diving to strike young reindeer, using their talons to bring them down.

Defense Strategies

Reindeer have evolved several ways to survive:

- **Watchful Eyes**: In herds, someone's always watching for danger
- **Speed:** At first sign of trouble, reindeer RUN and can keep running for miles
- **Protective Circles:** Cows with calves will form groups with babies in the middle
- **Fighting Back:** Cornered adults use sharp hooves to kick and antlers to stab
- **Swimming Away**: Reindeer can swim to escape land predators
- **Smart Locations**: Calves are born in areas with fewer predators

Despite these defenses, reindeer losses to predators are high each year, especially among calves. Yet this natural cycle keeps Arctic ecosystems in balance. Predators depend on reindeer for survival, and in turn, reindeer populations are kept healthy by removing the weak and sick—a balance that has shaped life in the frozen north for thousands of years.

Challenges and Threats

Reindeer have survived for thousands of years, but today face growing challenges, many of them caused by humans and climate change. Some herds have shrunk by 90% or more in recent decades. Across the Arctic and subarctic, their world is shifting in ways that make survival increasingly difficult.

Climate change is the biggest threat. The Arctic warms twice as fast as the rest of Earth, creating cascading problems. Winter rain creates ice layers reindeer can't dig through, blocking food and causing mass starvation. Warming changes which plants grow where, with shrubs spreading into open tundra faster than reindeer can adapt. Earlier snow melt creates dangerous mismatches—babies born before or after peak food availability. Warmer temperatures mean worse mosquito harassment, and thinner river ice has caused mass drownings.

Another increasing threat is habitat loss and fragmentation. Expanding roads, pipelines, logging, and mining projects cut through ancient migration routes, separating herds from feeding and calving grounds. Noise and human presence disrupt the traditional movement patterns reindeer have followed for thousands of years. Overhunting and vehicle collisions have also increased as human development expands into reindeer territory.

Diseases and parasites are spreading northward with rising temperatures, exposing herds to new infections. Domesticated reindeer risk further disease transmission from livestock, weakening wild populations. For Indigenous communities who depend on reindeer for food, transport, and culture, these losses are deeply felt. Rare subspecies such as the Peary and Svalbard caribou are especially at risk due to their small, isolated populations.

All these pressures combine to push struggling populations toward collapse. Many groups are working to protect reindeer through protected corridors, monitoring programs, partnerships with Indigenous communities, and international cooperation. The future depends on protecting habitat and addressing climate change to preserve an ecosystem that has supported these animals for millennia.

How You Can Protect Reindeer

💡 **Save energy.** Turn off lights and electronics—less energy use means less global warming.

♻️ **Recycle and reuse**. Cut down on plastic and waste to protect habitats.

📔 **Learn and share.** Teach friends and family about reindeer and Arctic wildlife.

🌲 **Support conservation**. Follow or donate to groups like WWF that protect Arctic species.

🔊 **Speak up.** Write, post, or present about why reindeer matter—your voice makes a difference!

Life Span and Population

In the wild, reindeer usually live 8–15 years, though a few tough survivors reach 20 or more. Calves face the hardest odds—freezing weather, hungry predators, and deep snow make their first winter a dangerous test. Adults have their own battles with wolves, harsh storms, and scarce food, but healthy herds can thrive for many seasons.

Counting reindeer isn't easy! These herds wander across thousands of miles of tundra, mountains, and forest, often in regions too remote for people to reach. Still, scientists have discovered alarming changes in recent decades.

In North America, caribou once roamed by the millions. Today, some herds remain strong while others have nearly vanished. The Porcupine herd of Alaska and Canada is one of the healthiest, still about 200,000 animals following their ancient routes each year. But others have suffered huge losses. The George River herd in Canada fell from 800,000 to just 8,000, and the Bathurst herd dropped from almost half a million to a few thousand. Woodland caribou in Canada are now listed as threatened, with shrinking forest habitat and growing human development adding pressure.

DID YOU KNOW?

Female reindeer live longer than males—partly because bulls use up so much energy fighting during the rut each autumn!

In Eurasia, the picture is brighter. Norway still has around 25,000 wild reindeer, and Russia is home to several hundred thousand wild herds plus millions of herded reindeer managed by Indigenous peoples. In Finland and Sweden, reindeer herding remains a vital tradition that helps protect the animals and the land they depend on.

Worldwide, wild and free-ranging reindeer total about five million, but their future is uncertain. Some populations are stable; others continue to decline. The International Union for Conservation of Nature (IUCN) lists reindeer as vulnerable, meaning they could become endangered if trends continue.

Reindeer are a lifeline for entire Arctic ecosystems and cultures. Protecting their migration routes, food sources, and the people who live alongside them means protecting the Arctic itself—one of the most fragile, beautiful places on Earth.

Conclusion

Throughout this book, we've explored what makes reindeer such extraordinary and resilient animals. From their long migrations across frozen tundra to their role in sustaining Arctic food webs, reindeer show how life can flourish even in Earth's harshest environments.

Reindeer are full of surprises. Though they may seem calm and gentle, they are powerful travelers built for endurance. Their thick coats, wide hooves, and keen senses allow them to survive fierce winters, frozen rivers, and months of darkness. Their strength and stamina have inspired people for centuries, from ancient hunters to modern scientists studying Arctic life.

These remarkable mammals also teach us about the importance of balance. As grazers, reindeer shape the tundra by trimming plants and spreading nutrients. As prey, they feed wolves, bears, and other predators. By protecting reindeer and their migration routes, we safeguard entire northern ecosystems that depend on their presence.

Today, some reindeer herds number in the hundreds of thousands, while others are dwindling under the pressures of climate change, habitat loss, and industrial expansion. Conservation programs, indigenous herding traditions, and scientific research are now working together to protect these vital animals and restore threatened populations.

As we look to the future, reindeer remind us of nature's endurance and interconnectedness. Their migrations link forests, mountains, and tundra across continents. Through their adaptability, ecological importance, and deep connection to human cultures, reindeer show that even in the coldest, most remote regions, life continues to persist with strength, beauty, and purpose.

Test Your Reindeer Knowledge!

Think you remember everything about these amazing Arctic travelers? See how many questions you can answer!

1. What's another name for reindeer in North America?
   A) Moose   B) Elk   C) Caribou   D) Bison

2. True or False: Both male and female reindeer grow antlers.

3. What special feature helps reindeer walk on snow and ice?
   A) Flat hooves   B) Furry feet   C) Sharp claws   D) Sticky pads

4. How far can some herds travel each year during migration?
   A) 300 miles   B) 3,000 miles   C) 30,000 miles   D) 300 feet

5. What color do reindeer eyes turn in winter?
   A) Blue   B) Brown   C) Green   D) Gray

6. Why do pregnant females keep their antlers through winter?
   A) To dig feeding craters   B) To defend food spots   C) For warmth
   D) Decoration

7. Reindeer mainly eat _____ during winter.
   A) Grass   B) Lichens   C) Berries   D) Tree bark

8. Which of these animals is a main predator of reindeer?
   A) Wolves   B) Penguins   C) Polar Bears   D) Owls

9. Why are reindeer sometimes called "ecosystem engineers"?
   A) They build dams like beavers
   B) Their migrations shape the land and help other species survive
   C) They make shelters under the snow
   D) They plant trees across the tundra

10. How long do wild reindeer usually live?
    A) 3–5 years   B) 8–15 years   C) 20–30 years   D) Over 40 years

Answer Key: 1-C, 2-True, 3-A, 4-B, 5-A, 6-B, 7-B, 8-A, 9-B, 10-B

STEM Challenge: Think Like a Scientist!

Reindeer survive some of the harshest conditions on Earth. Try these fun, hands-on science experiments to discover how their bodies and behaviors help them live in the frozen Arctic!

Arctic Insulation Test

Topic: Adaptation & Heat Transfer

You'll Need:
2 small bowls, 2 ice cubes, 2 resealable plastic bags, cotton balls (or wool, paper towels), timer.

What to Do:

1. Place one ice cube inside a plastic bag surrounded by cotton (this is your "reindeer fur").
2. Place another ice cube in a plain bag with no insulation.
3. Set both in bowls to catch drips.
4. Check every 5 minutes—which ice cube melts slower?

What You'll Learn:
Reindeer fur is made of hollow hairs that trap air, keeping warmth in and cold out—just like the cotton around your ice cube!

Snowshoe Science

Topic: Physics & Engineering

You'll Need:
Tray of flour or sugar, cardboard, tape, small weights (coins or marbles).

What to Do:

1. Cut two pieces of cardboard: one small (3x3 inches) and one large (6x6 inches).
2. Tape a coin or marble to each piece of cardboard.
3. Press both "hooves" gently into the flour or sand.
4. Compare the depth of the prints—what do you notice?

What You'll Learn:
Wide reindeer hooves act like snowshoes, spreading out their weight so they don't sink into soft snow or mud.

Word Search

```
M E I N O I T A R G I M M R Z
R C O J O I N W F V H E R D S
R U M I N A N T S S E V O O H
M N N A W W Z D R T A V S P O
E H C N B O S P I S U L D R O
T D I T U L X R V G A N M U F
S H T L L V F E S M E W D F V
Y I C E L E M D M Q S N F R A
S F R R S S Q A Z S P F O M A
O C A S V Y M T V O Y E Y U E
C W L H I P U O B I R A C R S
E U C I H F N R X X O C O Y X
L F E V M E F S V W T V O V O
N H O Y H A T A F B I Z A W R
V B H C B V T T Z B R V K Z S
A E I T H K E E R X R B A Q G
M L R E I N D E E R E B J D L
O O P Z C X H Z O B T O M K R
```

Antlers	Fur	Migration
Arctic	Herbivore	Predators
Bulls	Herds	Reindeer
Caribou	Hooves	Ruminants
Climate	Indigenous	Territory
Cows	Lichen	Tundra
Ecosystem	Mammals	Wolves

Glossary

adaptations – special features or behaviors that help a plant or animal survive in its home

Arctic – the northernmost region of Earth, around the North Pole, where it's extremely cold and much of the ground stays frozen year-round

boreal forest – northern evergreen forests found just south of the tundra; also called taiga

calving grounds – traditional areas where female reindeer gather to give birth to their young

caribou – the North American name for wild reindeer

ecosystem – all the living things and their environment in an area

herd – a group of animals that live and travel together

herbivores – animals that eat only plants

Indigenous – people who are the original inhabitants of a region; native peoples who have lived in an area for thousands of years before others arrived

lichen – crusty or fuzzy organism made of algae and fungi living together; an important food source in the Arctic

migration – regular movement from one place to another, usually with the seasons

predators – animals which hunt other animals for food

ruminants – animals that chew their cud (regurgitate and re-chew food)

rut – the breeding season for deer and related animals

tundra – flat, treeless Arctic land where the ground stays permanently frozen beneath the surface

Resources and References

Want to learn more about reindeer and the Arctic? Check out these trusted books, websites, and organizations that explore wildlife, science, and conservation across the frozen north.

Books

Caribou and Reindeer by Dorothy Hinshaw Patent (Clarion Books) — A detailed look at these amazing Arctic travelers.

Animals of the Arctic by Peggy Thomas (Capstone Press) — Discover how Arctic species adapt to life in extreme cold.

The Big Book of Animals of the North by Susan Hughes (Kids Can Press) — Beautiful photos and facts about northern wildlife.

Websites

National Geographic Kids – Reindeer Facts
kids.nationalgeographic.com/animals/reindeer
Learn about reindeer behavior, migrations, and conservation.

World Wildlife Fund (WWF)
worldwildlife.org/species/caribou
Explore global efforts to protect Arctic wildlife and habitats.

Arctic Portal
arcticportal.org
Maps, news, and scientific updates about Arctic climate and ecosystems.

Canadian Wildlife Federation – Caribou
cwf-fcf.org
See how Canada tracks and supports its caribou populations.

For Young Scientists

NASA Earth Observatory – The Changing Arctic
earthobservatory.nasa.gov
Satellite images showing how ice and snow are changing over time.

Polar Bears International – Arctic Ecosystems
polarbearsinternational.org
Learn how Arctic animals—including reindeer—depend on each other.

Keep Exploring!

If you enjoyed learning about reindeer, explore other titles in the This Incredible Planet series to discover more amazing animals—from sea turtles to penguins to elephants—and the habitats they call home.

Index

A
adaptations, 12
antlers, 8, 11, 12, 16
Arctic, 7, 11, 27, 31
Asia, 11, 32

B
barren-ground caribou, 11
bears, 28
birth, 23
boreal forest, 11
breeding, 20, 23
bulls, 8, 23

C
calves, 12, 23, 24
calving grounds, 11
caribou, 7, 11
climate change, 7, 15, 31
communication, 16, 19
cows, 8, 23

D
diet, 15
diseases, 31

E
ecosystems, 27, 28
environment, 11, 27
Europe, 11, 32
eyes, 7, 8, 12

F
females, 8, 16, 23
food sources, 15
fur, 7, 8, 12

G
golden eagles, 28

H
habitat loss, 31
herbivores, 15
herds, 7, 16, 31
hooves, 8, 12, 16
human activities, 7, 31

I
Indigenous people, 7, 11, 31

L
lichen, 11, 15
life span, 32

lynx, 28

M
males, 8, 16, 23
mating, 23
migration, 7, 16, 18, 19
moose, 7
mosquitoes, 20
mountain caribou, 11

N
North America, 11, 32
nose, 12

P
parasites, 31
Peary caribou, 11, 31
physical characteristics, 8, 12
population, 31, 32
Porcupine caribou, 16, 32
predators, 16, 27, 28

R
reproduction, 23
ruminants, 15
rut, 23

S
senses, 8, 15
size, 8
sleep, 20
smell, 15
social life, 16
stomach, 15
summer, 11, 15, 16, 20
Svalbard caribou, 31

T
tendons, 12, 19
territory, 11
threats, 31
tundra, 11, 27

U
ultraviolet light, 8, 12

W
winter, 11, 15, 16, 20
wolverines, 28
wolves, 28
woodland caribou, 11, 32

www.ingramcontent.com/pod-product-compliance
Lightning Source LLC
Chambersburg PA
CBHW040224040426
42333CB00051B/3438